LARAMIE JR. HIGH IMC
WITHDRAWN

How Electricity Is Made

Text   C.L. Boltz
Design   Arthur Lockwood

Contents

page 2 What is electricity?
      4 Discovering the electron
      6 How electricity is made
      8 The power station
     10 Power from steam
     12 The generator
     14 Power from water
     16 Natural energy
     18 Nuclear power
     20 Transmission and distribution
     22 Power control
     24 Electricity in the home
     26 Cells and batteries
     28 Planning power for the future
     29 Important dates
     30 Glossary
     31 The energy that we use
         Who makes electricity?
     32 Index and acknowledgements

LARAMIE JR. HIGH IMC
1355 N. 22nd
LARAMIE, WY 82070

Facts On File Publications
New York, New York • Bicester, England

What is electricity?

Everyone has some idea of what is meant by 'energy'. We use physical energy when we run. Heat energy comes from the sun. Mechanical energy is used in machines. Light energy comes from lamps. Energy, in other words, is the fundamental stuff of all life and activity.

The most flexible and controllable form of energy is electricity, and we use it in many different ways – to cook a meal, freeze food, heat a room, provide light, power a train, communicate with men on the moon or with space vehicles even further away.

Lightning is the most dramatic form of natural electricity. It is caused when energy builds up inside thunder clouds and is released in a flash of electric current.

Though there are some natural forms of electricity, it is difficult to make use of them. The most spectacular example is the thunderstorm. This is caused by a massive build-up of very hot summer air rising rapidly and hitting the water-drops in the clouds. When the rising air collides with falling drops of water an electric **charge** is formed. The amount of energy created can be more than that of a high-explosive bomb. When the energy is great enough it turns the air into a **conductor**. A flash of current – or lightning – then follows, lasting perhaps less than a millionth of a second. The air expands as the hot lightning passes through it, then cools and contracts. The surrounding air rushes in and you hear thunder. The flash of lightning may reach the ground and strike a tree, or even a person, though this is rare. High buildings are protected from lightning by copper strips, or conductors, which run from the top of the building to the earth.

Certain animals, such as fire-flies and glow-worms, can create electricity from biological and chemical energy. The electric eel has many small cells along its body which produce enough electricity to stun or kill its prey. But, like thunderstorms, these natural forms of electricity cannot be harnessed to produce the power which we need in everyday life: it can be made only with the help of other forms of energy, such as oil, coal and water.

Glow-worms are beetles which create a yellowish-green light by a chemical process in their bodies. They can flash their lights in a special rhythm to attract each other.

Static electricity

The electricity produced in lightning is called **static** electricity because it does not flow through cables but instead builds up a charge. Hundreds of years ago it was discovered that if amber (fossilized pieces of resin from long-extinct trees) was rubbed with fur or a dry cloth, a tiny charge of static electricity was created which attracted little bits of paper. You can try the same experiment by rubbing a pen or a rod of non-conducting material such as glass or plastic against your coat sleeve. The Greeks called amber 'electron'. It is from this word that we get 'electricity'.

Rub a plastic pen against your sleeve and hold it above some small pieces of paper. The bits of paper will jump to the pen and stick to it. They are attracted by the static electricity in the pen. After a minute or two the electricity will drain out of the pen into your body and the pieces of paper will fall off.

Making use of electrostatics

For centuries it was not possible to put static electricity to any practical use. In recent years, however, electrostatic machines driven by electric motors have been invented for a variety of purposes, such as spraying paint on cars. Extremely high charges of static electricity can be built up in a **Van de Graaff machine** for use in nuclear physics. A continuous belt of charged brushes transfers electricity to a storing device at the top of the machine, thus building up a very high charge. Daresbury Laboratory in north-west England has the largest Van de Graaff machine in the world. Though it can produce up to 20 million volts, it still operates on the same simple principle as that of a flash of lightning or the rubbing of fur on amber.

An inside view of the Daresbury Laboratory's Van de Graaff machine which is used for experiments in nuclear physics. The central terminal can be charged up to levels of 20 million volts, using a special charging system based on the same principle of static electricity as that in the pen and paper experiment.

Discovering the electron

Static electricity is created by rubbing two different insulating materials together. This charge can then be transferred to another insulating material by touching it with the charged object. The charge can also be transferred by bringing the charged object *near* another material. This method of transferring a charge is called **induction**. The material is charged, but in a different way. For example, after amber has been rubbed with fur the two attract each other and the fur hairs stand on end. But if the charge on the amber is transferred to two corks they repel each other. Names had to be invented to describe this difference, so one charge was called **positive** and the other **negative**. The amber is said to be charged negatively and the fur positively. The objects with similar charges repel each other, whereas two objects with opposite charges attract each other.

The world consists of **matter**, which can be defined as anything that has weight and takes up space. Scientists used to believe that the smallest particle of matter was an **atom**. Then, in England in 1897, Sir Joseph Thomson made the dramatic discovery that there is a particle of

When amber and fur are rubbed together, static electricity is created. If the amber is held above the fur, the hairs will stand on end, showing that the two materials are now attracted to each other. If the charged piece of amber then touches against another insulating material, such as cork, its electric charge will transfer to the cork.

You would expect two charged corks to be attracted to each other, like the amber and fur, but, instead, they will move away from each other. The charge on the amber is therefore different from the charge on the fur. To describe this difference, one charge is called positive and the other charge is called negative.

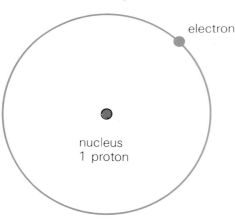

Hydrogen atom

All matter is made up of atoms. But atoms must have differences, otherwise everything would look and feel exactly the same! Each atom consists of a nucleus containing protons (with a positive charge) and neutrons (no charge). Electrons (with a negative charge) circle round the nucleus like planets round the sun. If one atom has a different number of electrons from another atom, it forms a different element.

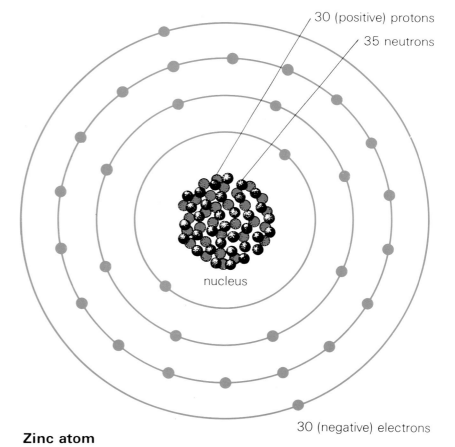

Zinc atom

matter even smaller than an atom, called an **electron**. This was the beginning of a new understanding of electricity.

In the early days scientists did not fully understand electric charges, but it is now known that when two dissimilar materials are rubbed together **electrons** are transferred. A negatively-charged object has too many electrons, but a positively-charged object does not have enough.

An atom consists of a central nucleus containing **neutrons** and **protons**. Electrons move around the nucleus, rather like planets moving around the sun. Electrons carry a negative charge and protons a positive charge. The atom of one element (a substance that cannot be broken down by chemical means into a simpler substance) differs from the atom of another element by its number of electrons.

Electrons

Electrons revolve in orbits around the nucleus. In the biggest atoms there may be as many as seven orbits. Generally the number of electrons equals the number of protons in the nucleus. This means that the negative charge on the electrons matches the positive charge on protons.

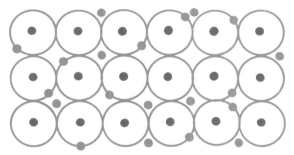

Conductor

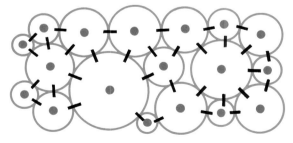

Insulator

Metal atoms have free electrons which wander from the outer orbit of one atom to the outer orbit of another. If a metal wire is connected to a battery, the free electrons start to flow from negative to positive. This flow is called electric current. Metals are good conductors of electricity.

Atoms in materials such as rubber or plastic have electrons which are tightly bound by the nucleus. When connected to a battery there are very few free electrons, so no current flows. Non-conducting materials are called 'insulators'.

Usually the force of attraction between positively charged protons and negatively charged electrons is so great that they are bound together. In metals, however, it is quite possible for an electron to leave its outer orbit and to travel to another atom where it starts orbiting its nucleus. When this happens it causes the atom to have one electron too many, so one of its electrons will move to another atom. The travelling electrons are called **free electrons**. They move at random in the metal, so that the positive and negative charges cancel each other out and no current flows. If electrical pressure is applied, by a battery for example, the free electrons flow in a controlled direction from negative to positive. This flow is called electric current. Metals, therefore, are mostly good conductors of electricity.

Materials such as rubber and plastic have electrons which are tightly bound by the nucleus. So if electrical pressure is applied there are very few electrons available to flow. Such materials are non-conductors, or **insulators**.

Materials which allow more electrons to flow than a good insulator but much fewer than a conductor, are called *semi-conductors* and are used in electronics, radios, etc.

Before electrons were discovered, electricity was thought to be a fluid which flowed from positive to negative. Although we now know that electricity really consists of electrons flowing from negative to positive, we still, from habit, wrongly talk about electric current flowing from positive to negative.

A section of underground cable used to carry power at 400 kv. The central copper conductor is covered with about 200 layers of insulating paper and enclosed in corrugated aluminium tubing to make it flexible enough to bend.

How electricity is made

If you fly over a city such as London, New York or Paris at night you will be fascinated by the pattern of lights sparkling out of the darkness, spreading out as far as you can see. As the plane descends you will be able to pick out the white and orange of street lighting, the warm glow from houses and blocks of flats, the slightly eerie brilliance of factories and office buildings, the endless streams of car headlights, and the different colours of the airport lights – green and red and purple and orange.

Can you imagine how much electricity is needed to produce so much light and heat? Few of us ever pause to wonder where it all comes from or how it is made.

The natural energy of rivers and oceans, the sun, the winds, and the heat beneath the earth's crust can all be harnessed to create electricity. But the most important fuels available at the present time are coal, oil, gas and diesel fuel, which have to be burned before they release the energy necessary for generating electricity. A source of energy which is now becoming increasingly important is **nuclear fission**.

In many areas of the world, fuels and water power are not readily available. This means that energy has to be transmitted over hundreds of miles. The creation, or generation, of this energy takes place in power stations, which burn fuel such as coal, oil or gas. The heat produced boils water, which turns into steam.

Nuclear power stations produce heat by a process called **nuclear fission** (described on pages 18/19). The steam is directed at a machine called a **turbine**, and causes it to rotate. If the power station is hydroelectric, the turbine is driven by water instead of steam. The turbine causes a machine called a **generator** to rotate, and the generator creates electric current.

nuclear reactor

Power stations burn coal or oil, or use nuclear fission, to produce heat. From the coal-bunkers fuel is carried on a conveyor belt to the coal pulverizing mill where it is ground to a powder as fine as flour. The powdered coal is blown into the boiler furnace where it creates enormous heat. Nuclear power stations use a process called 'fission' to heat carbon dioxide gas in the reactor. The heat created by burning coal, oil or nuclear fission raises the temperature of the water as it circulates in the boiler-tubes, and turns it into steam.

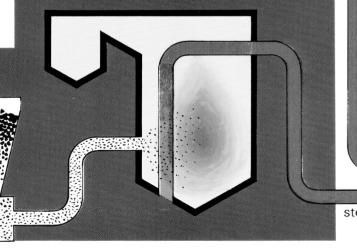

coal **pulverizing mill** **boiler**

City lights – an aerial view of Los Angeles. Where does all the electricity come from and how is it made?

The steam travels along pipes to the turbine. It hits the turbine blades and makes them spin. The spinning turbine-shaft drives the generator, which produces electricity. Steam is turned back into water in the condenser, where it can be recycled. The electricity generated flows through cables to a transformer. The transformer changes the electricity from a low voltage to a very high voltage, as this is the best way of transmitting electricity over long distances. The electricity is fed to a grid system of overhead lines and underground cables. Transformers in other parts of the country change the high voltage to lower voltages, for use in factories and homes.

The electric current is passed to a **transformer** which increases it to a very high voltage so that it can be transmitted over very long distances without losing too much power. The current flows through thick wires, supported by steel pylons, to substations, where it is reduced to levels suitable for use in houses and factories. In cities and built-up areas the power cables are usually underground.

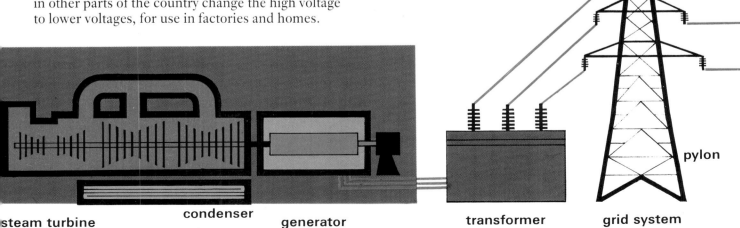

steam turbine   condenser   generator   transformer   grid system   pylon

7

The power station

The currents and voltages which we use in everyday life are fairly small. Even if we switch on all our household appliances at the same time we never use more than a few kilowatts (kW) of power (1 kilowatt equals a thousand watts). But the power generated in a power station is measured in megawatts (MW; 1 megawatt equals a million watts). A power station such as Ratcliffe in England is capable of generating 2000 megawatts.

The most impressive aspect of many coal-fired power stations is the series of tall towers with concave sides. These are cooling towers. In producing energy for electricity, water is required to cool the system. The water removes the heat and is then dropped down inside the towers so that it can evaporate and cool for re-use. The steam created emerges from the tops of the towers.

The Ratcliffe coal-fired power station near Nottingham generates 2000 megawatts of power. The coal stock area is in the foreground. Coal is taken by conveyor belt into the boiler house building (centre). The building on the left, joined to the boiler house, contains the turbines. The low, separated building further to the left is the substation from which pylons transmit electricity. The water for the eight cooling towers comes from a nearby river. The 198 metre (650ft) chimney discharges the gases produced by burning coal.

A pressurised water-reactor (PWR) nuclear power station at Callaway, Fulton, Missouri, USA, during construction.

In addition to cooling towers the coal-fired power station also has a tall chimney. Plates inside this chimney are charged with static electricity. They attract grit and small particles and so remove them from the waste products produced by combustion. But smoke and gases, such as sulphur dioxide, remain and are discharged into the atmosphere. These emissions could be contributing to so-called 'acid rain' which can damage the biology of rivers, lakes and forests.

The buildings of a coal-fired power station house the boiler, turbines, generators, and the substation (where the electrical switch gear is situated). There is also a large coal storage area.

A nuclear power station, although cleaner than a coal-fired station, is perhaps less impressive in appearance and does not give quite the same feeling of massive power. Except for the earliest stations, there are no cooling towers or chimneys. This is because water for cooling is taken from the sea; the supply is unlimited and it does not need to be re-used. The taller buildings contain the reactors, and the turbine hall is usually separate.

Power stations do not *have* to be very large. In remote, less populated areas of the world enough electricity can be made from a small generator run by a diesel engine. Some big manufacturers and large hospitals also have their own diesel or petrol-run generators for use in emergency during a power failure.

The over-all efficiency of power stations is not very high, and some 60 per cent of the original energy is wasted when the hot water is cooled. Various schemes for using the hot water have been implemented – for instance heating buildings or growing hothouse vegetables – but further research is needed before such ideas can be put into practice on a world-wide scale.

> We measure electrical pressure in **volts**. The rate of flow, or current, is measured in **amperes**. If we multiply amperes by volts for any usage we get power, measured in **watts**. If we compare different metals to see how efficiently they conduct current we use a unit called an **ohm**, which measures the metals' resistance to current.

Power from steam

Between 80 per cent and 90 per cent of all electrical power in the USA and UK is created from the energy in steam. The steam for power stations is at high pressure, ranging from about 40 times to almost 200 times that of the atmosphere. The steam drives a turbine, which in turn drives the rotor of **a generator**.

To make the steam, heat is needed, and it is in the type of fuel used that power stations differ. Coal-fired stations are the commonest in both the USA and UK. Other fuels, even peat, are used in some countries. In a nuclear power station the nuclear reactor is the boiler.

The principle of a turbine is easy to understand: it is based on the windmill, in which the wind turns the vanes, or blades, which then drive a wheel to grind corn or pump up water.

As the steam is very hot, the blades must be of metal that will withstand great heat, without melting, softening or warping. The diameter of the circle on which the blades are mounted varies according to the pressure of the steam. The blades used for high-pressure steam are of smaller diameter than those used for lower-pressure steam.

Large turbines
The turbine, condenser and pipes are enclosed in cylinders to prevent steam from escaping. The high-pressure steam from the boiler goes first to a series of high-pressure smaller blades where energy is extracted. The steam therefore loses pressure and temperature, and expands. It is then passed back to a section of the boiler for re-heating. From there it goes to the part of the turbine designed for intermediate pressure and, after that, to the larger blades of the low-pressure area.

When the steam has lost as much energy as can be efficiently used, it is still very hot, and it passes to a condenser where it can be cooled and so condensed into water. This water returns to the boiler to be re-heated to high-pressure steam. None of the water is lost.

A turbine-generator unit being installed at a coal-fired power station. The diameter of the blades varies according to the pressure of the steam. The blades used for high-pressure steam are smaller in diameter than those used for low-pressure steam.

This huge bladed rotor is part of a steam turbine-generator used for generating electricity.

Below: High-pressure steam from the boiler enters the high-pressure section of the turbine and passes through the sets of blades. Energy is extracted, and the steam returns to the boiler to be reheated. The steam then enters the medium-pressure section and is passed through its blades to low-pressure blades. The energy extracted turns the turbine shaft which turns the generator. When all the energy possible has been extracted from the steam it is expelled to the condenser, converted to water, and returned to the boiler.

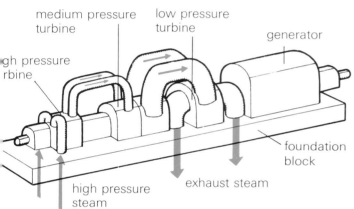

The condenser cooling water requires a separate supply of water. Where the cooling water for power stations is drawn from large rivers or from the sea it can be returned directly to the source after use. Power stations situated on smaller rivers or inland do not have such vast water resources available, so the cooling water is passed through cooling towers (where its heat is removed by evaporation) and re-used.

It can be seen that a turbine used in large power stations is a very complex piece of engineering, with the cooling system taking up as much space as the turbine itself.

The turbine hall at the 2000 MW coal-fired Cottam power station in Nottinghamshire, England. It contains four 500 MW turbo-generator units.

The generator

Thousands of years ago it was discovered that lodestone (now known to be a form of iron ore) could attract pieces of iron. If the lodestone, or magnet, was suspended so that it could swing freely, one end would always point towards the north and the other end towards the south. The two ends were called **poles**.

If you take a magnet in the form of a bar, place it on a flat surface, and sprinkle iron filings around it, the filings will form a pattern which will indicate the magnetic force, or field, surrounding the magnet. The pattern will consist of lines called **lines of force**.

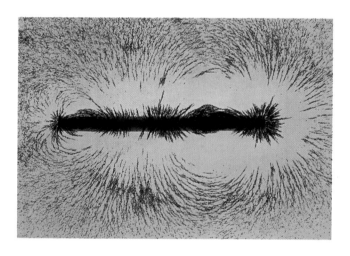

Magnetic fields can be used to produce electricity. The first person in history to discover the connection was a Danish scientist, Hans Christian Oerstad, in 1820. As a result of his work and that of André-Marie Ampère in France, it was found that a magnetic field could be created by means of an electric current flowing through a wire. Bend a long coil of wire into loops and pass a current through it and the coil becomes, for the time being, a bar magnet. This is called an **electromagnet**. Today they are a common part of electric motors and generators.

Alternating current

In England in 1831 Michael Faraday discovered how to generate electricity from magnetism, and so founded the world's electrical industry. A long coil of wire is joined to a galvanometer, or an ammeter, which are instruments for measuring electric current. If a bar magnet is pushed *into* the coil, the instrument needle will move, indicating that an electric current is flowing. If the magnet is pulled *out of* the coil the needle will move in the opposite direction. If the magnet is left quite still inside the coil, the instrument needle will not move at all. To make a current there must be movement. If the magnet is pushed *quickly* in and out of the coil the needle will move back and forth, indicating that current is flowing all the time. This current periodically changes direction because it is alternately cutting the magnetic lines of force in opposite directions. It is called **alternating current** or AC.

In this simple experiment the current created is very small. The easiest way to produce a useful amount of current is to cut the magnetic lines of force by rotating a magnet inside a coil or by rotating a coil inside a magnetic field.

In modern generators the conductors are fixed, and the magnetic field rotates inside. The stationary part is known as the **stator** and the magnetic field is created in the **rotor**. Current from the stator is fed to conductors called **bus-bars** and from there to the **transformers**.

The generators supplying alternating current are called **alternators** and they supply AC for transmission to all users of electricity. The time taken for one complete alternation of current is called a **cycle**. The number of cycles per second is called the **frequency**. In the UK this is standardized at 50 cycles per second and in the USA at 60.

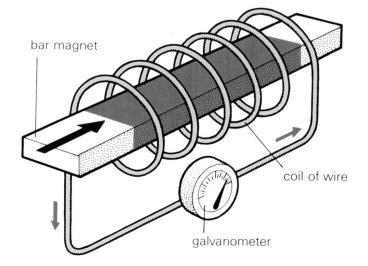

If a magnet is pushed quickly in and out of a coil of wire that is joined to a galvanometer, the needle moves back and forth, indicating a flow of electric current.

Direct current

Direct current, or DC, flows in the same direction all the time. A DC generator resembles an alternator except that it has a device which mechanically changes the AC generated to DC. Argument about which is more efficient, AC or DC, has gone on for years.

The advantage of AC is that it can be easily changed from one voltage to another by means of a transformer. It can then be transmitted at high voltage by overhead lines over very long distances without a significant loss of power. If, however, it is transmitted by underground or submarine cable there is considerable loss of power.

DC can be transmitted over long distances by cable. It can also drive motors for transport. All London's underground railway trains rely on DC motors, as does the New York subway system. It is also widely used for main-line trains.

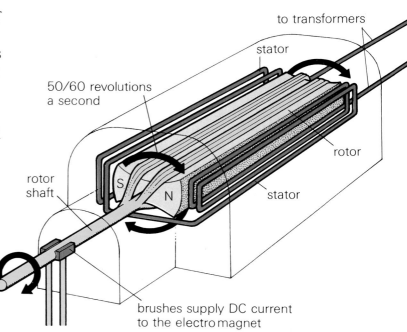

The revolving field AC generator is the type most widely used. The revolving part is called the rotor and is usually driven round by a steam turbine (the super-heated steam being produced by burning coal, oil, or nuclear fuel). Direct current from a separate source passes by means of brushes to the rotor to create a rotating magnetic field. The magnetic field extends through the surrounding stationary coils in the stator, and as the rotor turns, alternating voltages are induced. These are 50 cycles per second (60 cycles in the United States). The current from the stator is fed to conductors and thence to transformers.

A power station generator under construction.

Power from water

The 221m (726ft) high Hoover Dam on the Colorado River created Lake Mead, the largest man-made reservoir in North America. The water passing through Hoover's 17 turbines generates about four billion kilowatt hours of low-cost energy a year – enough for 500,000 homes.

Hydroelectricity is generated by constructing a dam across a river and building a power station below the reservoir. The distance can vary from a few hundred to a few thousand feet. When valves are opened, the water from the dam falls through large pipes to the turbines. The pressure of the water spins the turbine blades. The turbines drive the generators and produce electricity.

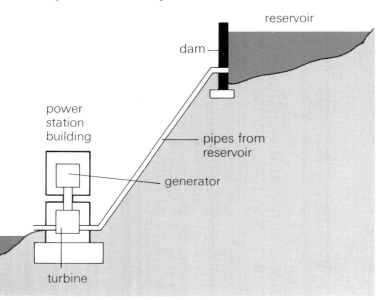

Flowing water is an obvious source of energy: consider the colossal power of Niagara Falls, for example, or the boundless energy of the ocean. The action of the wind on the sea creates waves powerful enough to sink ships and to destroy concrete breakwaters. Tides can vary as much as 18 metres (60 feet) between high and low water, and they contain enough energy to erode coasts.

People began to harness these unlimited sources of natural energy many centuries ago, with water wheels for grinding corn. As time went on, more powerful water wheels were invented to operate industrial machinery, and eventually engineers began to see the possibilities of using water to generate electricity.

Hydroelectric power stations

Flowing water can be used for turning turbines on a similar principle to water wheels. A 'head' of water is created by damming a river. A power station is built below the dam – the difference in height varying from a few hundred to a few thousand metres. Valves are opened, and water is allowed to fall from the dam through large pipes to the turbines. The turbines then drive the generators.

Hydroelectric power stations in the UK generate only a fairly small output of electricity, but in the USA there are huge stations, such as those of the Tennessee Valley Authority. Electricity produced from such stations amounts to some 15 per cent of the total energy used over the whole country.

Countries with fast-flowing rivers, such as Sweden and Norway, rely heavily on hydroelectricity; and the mighty rivers of Brazil and China make gigantic hydroelectric projects possible. China's hydroelectric power stations produce a total of 35,000MW. Brazil is building the world's largest hydroelectric scheme, at Itaipu, which will generate over 12,000MW.

Tides and waves

The world's best-known tidal power station is at the mouth of the river Rance in France. In this station 24 **turbo-alternators** operate as pumps or as turbines. Each turbo-alternator faces downstream so that it can generate electricity from the incoming tide. Water is pumped into a reservoir, and when the tide falls the turbines are

The world's largest and best-known tidal power station stretches across the mouth of the river Rance in France. Electricity is generated from the incoming tide and, at the same time, water is pumped into a reservoir. When the tide falls, electricity can be generated by releasing water from the reservoir.

Below: Dinorwig in Wales is Europe's largest pumped storage power station. This view of the vast machine hall under construction gives some idea of the engineering skills involved in building it. The machine hall holds the turbines and generators – it is twice the size of a football pitch and the height of a 16-storey building.

reversed and water is released from the reservoir. There are a few small tidal stations in the USSR but, despite extensive research, tide generation has not been put into use on a very large scale anywhere in the world.

Recently, considerable progress has been made in countries such as the UK, Japan, Israel and the USA on various schemes for extracting energy from waves. It is possible that in the near future at least one of them will be successful on a practical basis.

Pumped storage

Electricity cannot be stored on a large scale, so when demand is lowest it is used for pumping water up to a storage lake. When demand increases, water from the lake is used to generate electricity. The largest pumped storage station in Europe is at Dinorwig in Wales. It is known as a 'peak-lopping' station and it helps to solve the problem of the variations in the supply and demand of electricity.

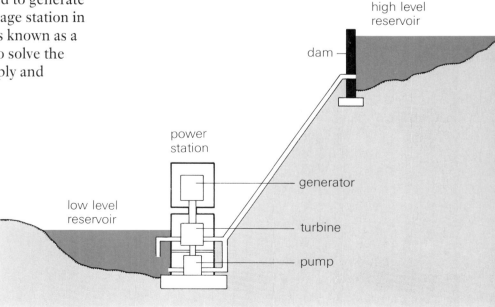

A pumped storage power station is built between two reservoirs. At night, when electricity is cheaper, water is pumped from the lower lake to the higher lake. During the day, electricity is generated when water from the higher lake is allowed to flow back through the turbines into the lower lake.

Natural energy

Wind

The natural energy of the wind is enormous, and for centuries its power has been used for windmills to grind corn and to pump water. In recent years new types of windmill have been designed for driving generators; the old-fashioned 'sails' have been replaced by aerodynamically-shaped propellers which are mounted on a rotating cylinder containing a generator. This is fixed to the top of a high steel tower. A computer in the control room at the base of the tower ensures that the rotor turns to face the wind. The electricity generated is led through a vertical cable to the bottom of the tower.

In the USA there has been considerable development of windmills, and 'wind farms' have been created. A landowner can plant crops around a number of wind-operated generators and sell the electricity to the local supplier. This can be a profitable addition to his income.

In the UK the Central Electricity Generating Board has built a small experimental wind-turbine at Carmarthen in Wales. It is capable of generating 200kW, which is enough electricity to supply a small village.

Another type of wind-turbine has also been developed recently in the UK. It has a vertical shaft, and blades which rotate horizontally. The generator is at the base of the tower, where it is easily serviced.

Wind-turbines work effectively only at certain wind speeds and will shut down automatically in storm winds. Therefore their power may not be available at moments of peak demand. Also, large groups of wind-turbines would disfigure the landscape and would create television and radio interference. A possible solution might be to construct windmills offshore, but in the UK this is unlikely to happen before the 1990s.

The potential of wind power as a valuable source of energy has in no way been fully exploited, and it will be interesting to see what progress is made in the next few years.

The Carmarthen Bay wind-turbine in South Wales. It begins generating electricity in a gentle breeze and closes down automatically in storm winds. It supplies up to 200 kW to the National Grid – equal to about 200 one-bar electric fires.

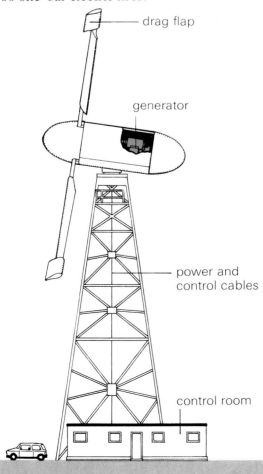

A three-bladed 24m (80ft) diameter steel rotor is mounted on a high steel tower. Drag flaps stop the rotor in storm winds. A computer in the control room monitors the machine's operation and makes sure that the rotor faces the wind. Power from the generator reaches the ground through a cable.

A wall of solar cells produces electricity, at Odeillo in France.

The Great Geyser in Iceland erupts, releasing geothermal energy that could be harnessed to make electricity. A power station would have to be built nearby and a borehole drilled through the rock to the natural fault in the earth's crust.

Sun

If we could efficiently harness even a fraction of the sun's energy we could solve all our energy needs. Sunlight can be converted directly into electricity by means of solar cells, usually made of silicon. However, the voltage and current produced by each cell are very limited, so a great many cells are needed to produce a useful wattage. Also, the cells are expensive to produce, though the cost is gradually decreasing.

Solar energy schemes have been introduced in parts of Africa and the USA. Solar panels are used in space and to heat water for homes in many parts of the world. Research is continuing in many countries, but a variety of problems still has to be solved before solar energy can meet our needs for electricity.

Geothermal energy

Hot, molten substances lie beneath the earth's crust. Where faults develop in the crust, geothermal energy is released through steam – as in geysers. This provides a valuable source of energy for countries such as Italy, Iceland and New Zealand. But as the steam has to be used near its source, there are many problems to be overcome before geothermal energy can be fully exploited on a large scale.

Nuclear power

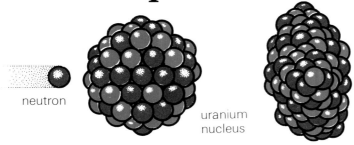

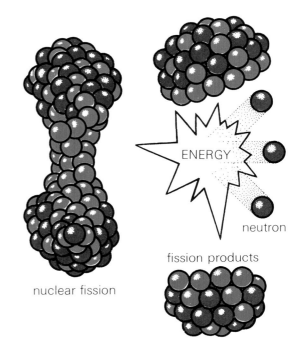

A nuclear power station is similar to a coal-fired station except that the heat used to make steam comes from energy released in a process called **nuclear fission**. Neutrons collide with uranium atoms, splitting them in two. A few neutrons escape at high speed and collide with other atoms, causing a chain reaction of splits (or fissions). Neutrons are more likely to cause fission in the usable part of natural uranium if they are slowed down by collision with a material known as a **moderator** – graphite in the British type of reactor, and water in the American type. In order to control the heat produced by fission a coolant is used – carbon dioxide in the British type, and water in the American. Because the American pressurised water-reactor (PWR) uses the same water both as a moderator and a coolant it is cheaper to build. It is, therefore, the most common type of nuclear power station.

Nuclear fission takes place when a slow-moving neutron strikes the nucleus of a uranium atom, causing it to split into two pieces which fly apart, generating heat. At the same time a few neutrons escape at high speed. When slowed down by a moderator they collide with other atoms, causing a chain reaction of splits (or fissions) and a steady generation of heat.

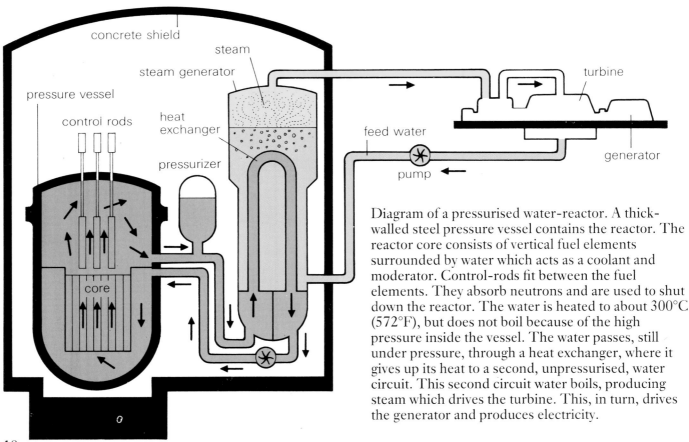

Diagram of a pressurised water-reactor. A thick-walled steel pressure vessel contains the reactor. The reactor core consists of vertical fuel elements surrounded by water which acts as a coolant and moderator. Control-rods fit between the fuel elements. They absorb neutrons and are used to shut down the reactor. The water is heated to about 300°C (572°F), but does not boil because of the high pressure inside the vessel. The water passes, still under pressure, through a heat exchanger, where it gives up its heat to a second, unpressurised, water circuit. This second circuit water boils, producing steam which drives the turbine. This, in turn, drives the generator and produces electricity.

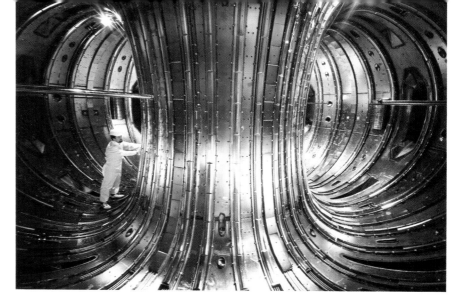

This is the interior of the JET vacuum chamber. JET is the Joint European Torus built in 1983 at Culham, Oxfordshire, England, as part of the European fusion research programme. The aim of this research is to produce reactions similar to those in the sun, but under controlled conditions for electricity production.

When the nuclear fuel has been used in a reactor for a long time it loses its ability to produce heat, and has to be replaced. The spent fuel contains unburned uranium and plutonium (created in the fission process) both of which can be re-used as fuel. It also has an amount of waste which is radioactive and could harm people exposed to it. Some is more highly radioactive than the rest. So safe disposal and storage are essential.

In a thermal reactor one tonne of natural uranium is equivalent in heat (if every bit could be used) to 20,000 tonnes of good coal. A new type of reactor which is at present being developed – the fast reactor – will produce many times this amount of energy per tonne.

Another potential source of power – **nuclear fusion** – is still at the research stage. The sun's energy comes from the fusing of atoms together under great heat. If we could bring about a similar fusion of atoms, such as deuterium (which can be obtained from seawater), by scientific methods, the energy released would provide almost unlimited power. However, it will be many years before we will know whether fusion is a practical and cheap source of electricity.

The interior of Hunterston 'B' nuclear power station in Scotland. Part of the reactor is shown in the foreground. In the background, the movable tower on rails is the charge/discharge machine used for inserting or withdrawing fuel and carrying out other maintenance work.

Transmission and distribution

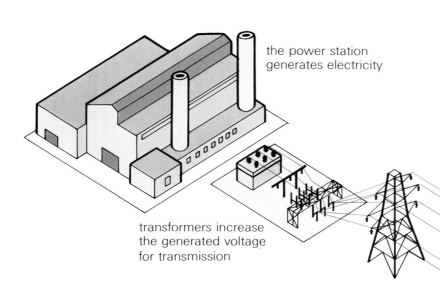

the power station generates electricity

transformers increase the generated voltage for transmission

This diagram shows how electricity generated by a power station reaches homes and factories by means of overhead lines, underground cables, and transformer stations.

overhead lines on pylons carry electricity across the countryside

The electricity generated by power stations has to be transmitted over an entire county or even a whole country. When current flows through a conductor it loses energy, depending on the resistance of the conductor and the amount of current flowing. In order to overcome this problem each generating station is connected to a transformer station where the voltage is increased: to 275,000 or 400,000 volts in the UK, and to comparably high voltages in the USA.

The principle of the **transformer** was developed by Faraday. He wound two separate coils of wire round an iron ring. When he passed a *changing* current through one coil by switching the current on and off, there was a corresponding current in the other coil. It was *induced* by the changing lines of force in and around the iron ring. If there were 20 turns on the first coil and 10 on the second, the voltage induced in the second coil was *half* that put on the first coil. Thus 100 volts could be transformed into 1000, or 1000 into 100, depending on the proportion of the turns.

From the transformer station, electricity is fed to a **grid system** which consists of overhead lines, or conductors, made from copper or aluminium, supported by huge steel pylons. The conductors are joined to the pylons by insulators. These very high voltage lines form the supergrid, and they carry the current to transformer stations where the voltage is reduced. Smaller pylons then carry the lines to regional transformer stations or grid supply points.

A transformer substation where very high voltages are reduced to lower levels. High voltage power lines enter and leave the transformer through the large brown insulators. The insulators keep the very high voltages away from the metal tank containing the transformer coils. Oil flows round the coils to keep them cool.

Distribution

Overhead lines and underground cables carry the reduced current from the regional transformer stations to local transformer stations which reduce the voltage still further, to levels suitable for industry and for houses. Domestic voltage in the UK is 240v and in the USA 120v.

In the UK this high-voltage transmission network is known as the National Grid. A grid network has one main advantage over systems where separate power stations supply local needs. The stations are generally sited away from the cities which they serve, for social reasons. It is essential for them to be near a river or the sea, and it is convenient for them to be near an energy source, such as an oil terminal or a coal field. Wherever they are situated, they can be linked by the grid network to the rest of the country.

The USA, which is a vastly bigger country than the UK, has networks designed to serve smaller areas. Some supplies are owned by a state, others by a community or by private enterprise. But there are interconnecting lines to enable one area to sell power to another.

High voltage powerlines are carried across country by pylons. The large insulators are designed to prevent electrical circuits from forming between the conductors and the ground.

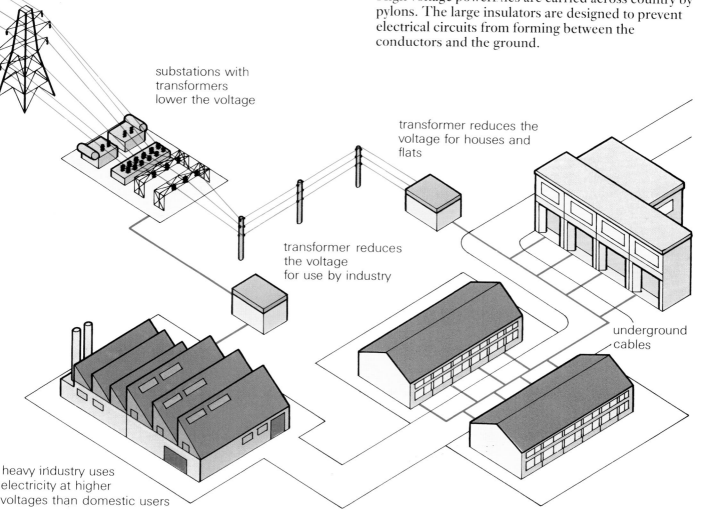

substations with transformers lower the voltage

transformer reduces the voltage for houses and flats

transformer reduces the voltage for use by industry

underground cables

heavy industry uses electricity at higher voltages than domestic users

A power station control room. Wall-displays summarise the minute-to-minute operation of the system. Computers monitor boiler and generator conditions.

Power control

Electricity has one disadvantage as a means of serving the needs of a community – it cannot be stored on a large scale. This can cause major problems, since the demand for electricity varies all the time. The peak times are in winter, in the early morning, as people get up and switch on kettles, lights, cookers and heaters, and in the evening when they get home from work. A popular television programme will suddenly increase demand, as millions of people switch on their receivers. Industrial use varies according to working hours and to the type of equipment in operation. Late at night, when most people are asleep, very little electricity is used.

Power stations work most efficiently and cheaply when they operate on full load for twenty-four hours a day. So, somehow, despite constant variation, a supply of electricity must meet the demand for it. This is achieved by shutting down and starting up power station generators as requirements vary.

Each power station has a control room where staff monitor the moment-to-moment variations

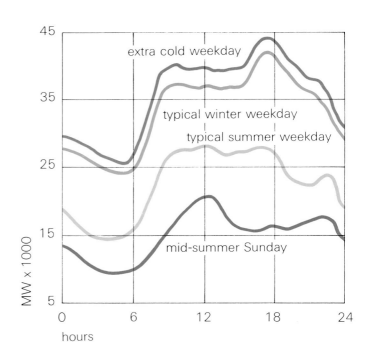

Graph showing the varying demand for electricity on two days in summer and two days in winter. During a weekday, demand is greatest during the working part, rising to a peak at about six o'clock. But on a Sunday the peak is at a quite different time.

in demand, and supervise the amount of electricity generated.

Power failures occur for a wide variety of reasons. Local faults, affecting a few houses or several streets, can be caused by workmen with power-drills cutting an underground cable; a dinghy mast hitting an overhead line; vandalism; a tree falling on a line; lightning; strong winds, and so on. Massive power failures like the famous New York blackout of 1965 can be caused by mechanical faults in the transmission and distribution system or by natural disasters such as floods, fire and ice storms. In most cases, however, the grid system makes it possible for power to be re-routed to the affected area from another part of the system. Engineers and technicians have to work round the clock until the fault is repaired.

The Central Electricity Generating Board in England operates the largest power system under single control anywhere in the world. The overall control centre for England and Wales is in London. The USA has areas of control rather than a single control centre.

Luminous displays, computer screens, teleprinters and large-scale maps enable staff to watch the minute-to-minute operation of the system – the power flows, voltage levels, switch positions, and so on. If the lines become overloaded, or the voltages exceed set limits, red lights flash on the control desks and on wall displays.

A well-planned distribution system ensures that users receive a constant voltage. In the UK, regulations specify that variation in voltage must not exceed 6 per cent of the specified value. In the USA the variation allowed is 5 per cent. This is known as the **tolerance**.

The control room at Dinorwig power station, Wales. This pumped-storage power station can begin generating electricity within one minute of instructions being received. It is ideal to cope with sudden demands for electricity which may occur, for example at the end of a popular television programme, when electric fires, kettles and lights are switched on in millions of homes.

Electricity in the home

In towns, electricity is transmitted to homes, offices, etc, by underground cable. In the countryside it is transmitted by overhead lines. The transformers in local substations reduce the voltage for domestic use to 240v in the UK, 120v in the USA and, usually, to 220v in Western Europe.

The electric cable comes into our homes via a sealed box containing the electricity authority's master fuses, to a meter which records the kilowatt hours that we use.

From this meter run insulated wires in metal or plastic tubes, hidden in the walls, under the floors or under the roof so that they cannot be touched or damaged. Switches and sockets enable us to plug in and use all the electrical gadgets in our homes.

Old-fashioned wiring systems have several sizes of plugs and sockets, and different thicknesses of wire. The safety device is a separate box with wires in it which will melt if too much current is used. The wires are called **fuses** and must be replaced with the right size of fuse wire. If a fuse melts it means that too much current is flowing and the wires are heating up.

A thin wire has more resistance than a thicker one, so it takes less current. If too much current is pushed through a wire it will heat up and may cause a fire. That is why care must be taken in houses with old-fashioned wiring systems. For example, a bedside lamp may be safely plugged into any **power point** because it only needs a low current, but a 1000-watt electric fire, cannot be plugged into a thin-wire circuit. In houses with modern wiring systems, any appliance up to about 3 kilowatts can be plugged into any power point.

Modern wiring systems are much safer. As well as the fuse box, each plug has a fuse cartridge which prevents too much current flowing to the individual appliance by disconnecting the **circuit**.

Understanding home wiring

Although there are obvious differences, it is easier to understand electricity if we compare it with water. If we want water to flow there must be pressure and there must be a pipe to carry the flow. As long as there is enough pressure, the wider the pipe the greater will be the flow; the

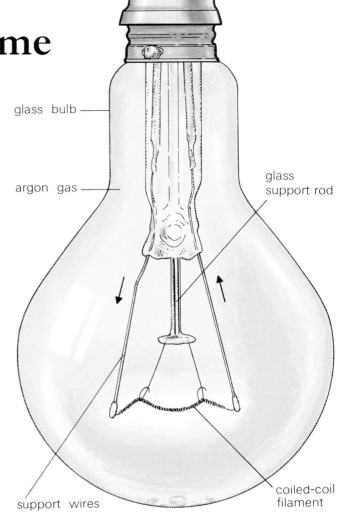

The very thin wire inside a light bulb resists the current passing through it, so the wire heats up and glows brightly. The more coils of wire in a bulb the more light is produced. The number on the bulb – 60w, 100w etc – tells how much power the bulb uses and how brightly it glows.

Diagram of a wall plug. Appliances have a wire fixed to the earth pin of the plug so that if a fault develops, the electricity will flow into the plug and then safely into the ground and not through the person holding the appliance! The fuse is another safety device which will disconnect the circuit if a dangerous amount of current is flowing. The correct fuse (3-amp, 5-amp or 13-amp) should always be fitted according to the kind of appliance. The neutral wire completes the circuit and allows electricity to flow back.

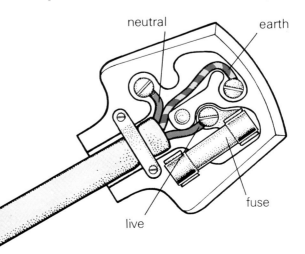

smaller the pipe the lesser will be the rate of flow. The same principle can be applied to electricity, though it flows through a metal wire or conductor instead of a pipe. Silver is the best conductor, but it is too expensive for common use. The next best is copper, and because it is cheaper it is universally used for wiring.

Just as water must be contained in a pipe, so electrical wiring must be insulated, or covered with rubber, enamel or plastic – often with a coating of cotton or some other fibre, and then the outer covering.

Electric current cannot start on its own. There must be a source of power such as a generator or battery; wires to conduct the current; and an appliance such as a light bulb. This arrangement is called a circuit. A switch allows the current to flow by completing the circuit, or stops the current by breaking the circuit.

Different appliances use different amounts of electricity. The cost, however, depends on the length of time in which we use the appliance. The unit of energy we pay for is the kilowatt-hour. The amount of energy used by a 1000-watt iron in one hour, for example, would last a 100-watt light bulb for ten hours.

Period of time that devices can be used for one unit i.e. one kilowatt-hour

Shower	10 to 20 minutes
Kettle	12 to 15 minutes
Dishwasher	20 minutes
Electric bar 3KW heater	20 minutes
Convector heater	30 minutes
Tumble drier	30 minutes
Iron	50 minutes
Vacuum cleaner	1½ to 2 hours
Oil-filled radiator	2 hours
Slow cooker	8 hours
Stereo system	8 to 10 hours
Television - colour	9 hours
100W bulb	10 hours
Fridge/freezer	12 hours
Sewing machine	13 hours
40W fluorescent light	20 hours
Television - b/w	20 hours
Refrigerator	24 hours

This picture shows some of the uses for electricity in your home. Appliances all use the same voltage, but different amounts of current. The unit of energy we pay for is the kilowatt-hour. The table shows the amount of electricity used by various appliances in your home.

Cells and batteries

Batteries are an essential part of our everyday lives. We use them in flashlights, bicycle lamps, calculators, radios, cameras and, of course, in motor cars. Yet few of us ever stop to wonder how they work or to ask why if car batteries can be recharged, flashlight batteries have to be thrown away.

A **cell** produces a voltage by chemical action. It would be possible to use many types of material to make a cell, but, in practice, cost and the possibility of corrosion or fumes limit the choice. A combination of cells will produce more voltage, or current, or both, and is called a **battery**. (The term is now wrongly used for both single cells and groups of cells.) Batteries can be classified into two types: **primary cells** which we use and then throw away, and **secondary cells** which are rechargeable.

Primary cells

The primary cell used most widely all over the world is the **zinc-carbon cell** (often called a 'dry cell') found in radios and flashlights. A zinc casing is filled with **electrolyte** made of ammonium chloride (which creates acid conditions), zinc chloride and manganese dioxide. This electrolyte is not really dry, yet it is not liquid. A porous separator made of flour and starch keeps the electrolyte from touching the zinc. A carbon rod runs through the centre of the

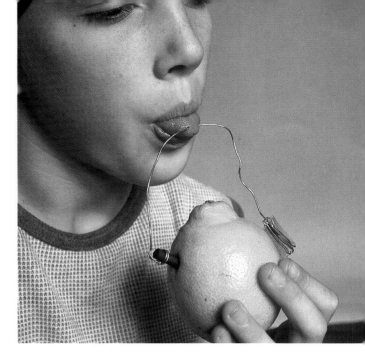

Cells can be made from many materials. Take a lemon and squeeze it, to break some of its inside tissue. Cut two slits in the skin. Take a carbon rod and a bit of the zinc casing from inside an old battery and push them through the slits into the lemon, making sure that they don't touch. Twist some copper wire around each, and hold the wires against your tongue. You should feel a very slight tingle, because your tongue will be conducting the electricity produced by the action of the acid in the lemon. Set up a second lemon in exactly the same way and double the power by joining the carbon rod to the zinc and the increased voltage will be noticeable.

cell. The carbon rod is the positive electrode, or pole, and the zinc casing is the negative. A chemical reaction in the electrolyte sends positive particles to one pole and negative particles to the other, so that when a wire is connected to the two electrodes, current flows. If a 'dry cell' is used and then left unused for a long time, chemical action will eventually destroy the zinc, and a messy fluid will ooze from it.

The zinc-carbon, and many other types of dry cell, will produce 1.5 volts. To obtain a bigger voltage, two or more cells are joined in *series* with positive and negative alternately or, to achieve more current, in *parallel* with positives together and negatives together. Nine-volt batteries, for example, consist of 6 small cells in series inside a metal case.

The increase in the need for power – to drive toys, camera flashes, etc – has brought about the development of several other cells. A zinc-carbon cell with zinc chloride instead of ammonium chloride in the electrolyte produces

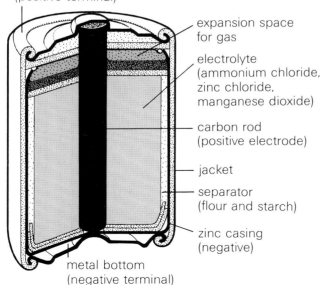

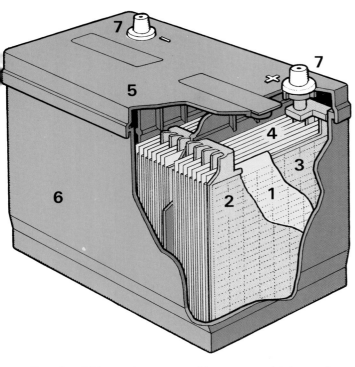

Lead-acid batteries are used in motor vehicles and can be recharged.
1 Grid – the metal framework for the electrodes.
2 Negative electrode packed with spongy lead.
3 Positive electrode filled with lead dioxide.
4 Separator – highly porous to allow acid to flow freely, but strong enough to prevent physical contact between the electrodes.
5 Lid – heat-welded to container to form an acid-tight seal.
6 Container – made in one piece, so it is very strong.
7 Battery terminals (negative and positive).

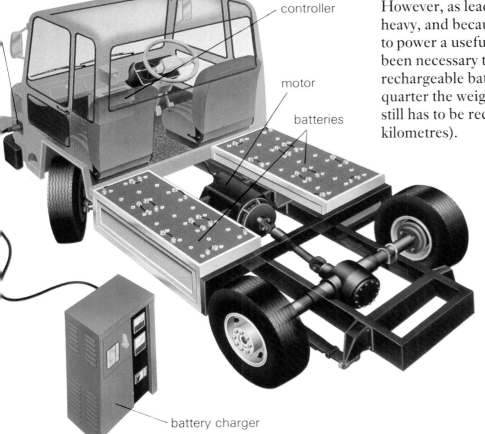

more current. An 'alkaline' cell will produce the same 1.5 voltage but it can, size for size, deliver up to 50 per cent more power. Its steel casing contains an electrolyte of potassium hydroxide. The negative electrode is powdered zinc mixed with mercury and the positive electrode is manganese dioxide and graphite. The separator is made of synthetic fibre.

The need for small 'button' cells for hearing aids, watches, cameras, etc, has led to the use of new and expensive materials such as mercuric oxide, silver oxide and lithium. Research and development continue as scientists strive to invent cells which will last longer and produce yet more current.

Secondary cells

The most familiar is the **lead-acid** battery used in motor cars. The electrodes are made of metal mesh. The spaces of the positive electrode are packed with lead peroxide paste and those of the negative with spongy lead. The electrolyte is diluted sulphuric acid. Each cell produces 2 volts, so a battery of 6 cells in series provides the 12 volts needed for a motor car. When the voltage reaches a low point it can be recharged by connecting the battery to a DC supply. The chemical reaction is then reversed, and the electrodes return to their original state.

Today there is a great interest in vehicles powered by electric motors because they are quiet, use no petrol and cause no pollution. However, as lead-acid batteries are large and heavy, and because at least 100 volts are needed to power a useful engine, extensive research has been necessary to discover other kinds of rechargeable batteries. The latest type is one quarter the weight of a lead-acid battery, but it still has to be recharged after about 60 miles (100 kilometres).

Vehicles powered by electric batteries are already performing a wide variety of duties – forklift trucks and maintenance vehicles in factories; milk floats, refuse collection vehicles and delivery vans on the roads. Electric vehicles are particularly useful in confined spaces where fumes and pollution could cause problems.

Planning power for the future

As the world's population increases and as developing countries become more industrial, the demand for electricity will increase. Supplies of oil, coal and gas are limited. It would be preferable, therefore, to save them for other important uses, such as fuelling cars and aircraft, heating our homes, and as raw materials for plastics, fertilizers, chemicals, etc. Also, burning these fuels can contribute to pollution.

Natural sources of energy such as the sun, winds and water will never run out – but we have not yet managed to harness them cheaply enough or on a large enough scale for practical use.

Nuclear power is cheaper and cleaner than coal and oil. Also, the world does not have an endless supply of fossil fuels, and we therefore will have to reduce our dependence on them.

When making plans for the future we have to consider cost, the continued availability of energy sources, and the risks involved. Making the wrong decisions now could leave countries dangerously short of electricity in the future.

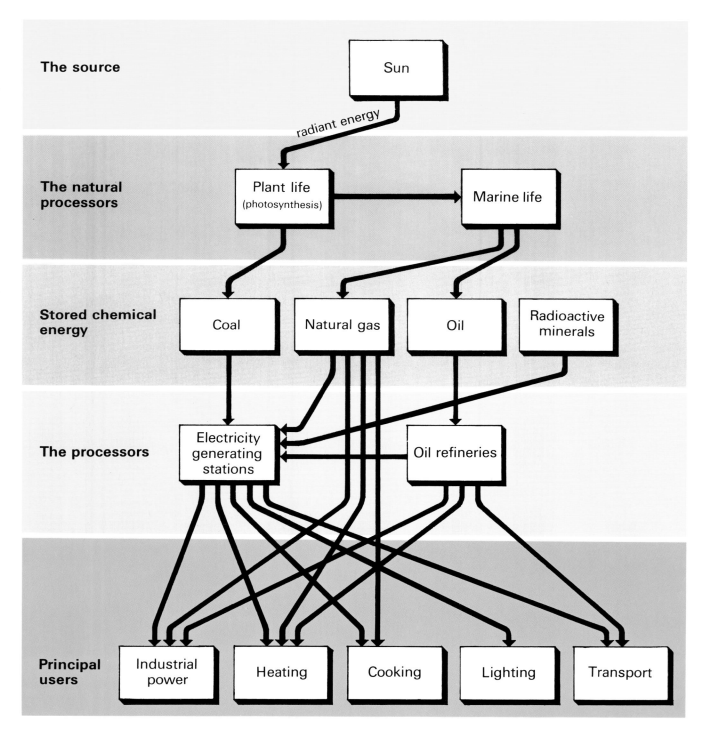

Important dates

Although the existence of static electricity caused by friction was known thousands of years ago, as was the magnetic quality of lodestone, it was not until AD 1600 that magnetism was treated seriously, and it was not until about AD 1800 that electricity became an important scientific study.

1600 William Gilbert (England) published his famous book, *De Magnete*, on magnetism and static electricity.

1752 Benjamin Franklin (USA) demonstrated that lightning was a form of electricity. He also brought into use the terms 'positive' and 'negative'.

1800 Alessandro Volta (Italy) made a battery which created continuous flowing electricity. It consisted of discs made from different metals separated by moistened absorbent material.

1820 The connection between magnetism and electricity was discovered by Hans Christian Oersted (Denmark). More comprehensive research into this connection was made by André-Marie Ampère (France).

1824 William Sturgeon (England) devised the first electromagnet.

1827 Georg Simon Ohm (Germany) outlined the first and most fundamental law of electrical circuitry. This is expressed today as: Current = Electrical pressure ÷ Resistance.

1831 Michael Faraday (England) discovered how to create electricity from a changing magnetic field, which was the basis of electrical generation. He also discovered the phenomenon of induction, which led to the invention of the transformer.

1832 The phenomenon of self-induction, or inductance, important in all circuit theory of alternating current, was announced by Joseph Henry (USA). The unit of inductance is still known as a 'Henry'.

1866 Georges Leclanché (France) made the famous primary cell which was in use for many years and which led to the zinc-carbon cell.

1878 The first rechargeable battery, invented by R. L. G. Planté (France), was exhibited in Paris. It was later improved by Sir Joseph Swan (England). With occasional improvements it has survived to this day.

1880 This period was notable for the AC v DC controversy in Europe and the USA. Among those supporting AC were Sebastian de Ferranti and S. P. Thompson in the UK, and George Westinghouse in the USA. Those against AC and for DC were Lord Kelvin, R.E.B. Crompton and John Hopkinson in the UK and Thomas A. Edison in the USA. A significant victory for AC came in 1893 when Westinghouse secured the contract to install alternators for the Niagara Falls scheme. In the meantime Ferranti was advocating high voltage for transmission in the UK.

1881 A golden year. Edison was building his first two power stations, one in London and one in New York. The first public supply was switched on in the small English town of Godalming. Soon after came the lighting of the British House of Commons; a New York City factory; and a number of ships, trains and theatre stages. The first AC generators were displayed at the Paris Exhibition.

1888 The first turbo-alternators were installed at a Newcastle power station in the UK. Sir Charles Parsons had developed the turbine after years of research.

1938–9 Hahn and Strassmann (Germany) demonstrated neutron-induced fission of uranium. Otto Frisch and Lise Meitner (Austria) showed that fission released energy.

1942 Enrico Fermi (USA) proved that controlled nuclear fission could take place, in the first-ever nuclear reactor, in Chicago.

1956 The world's first nuclear power station of commercial size was opened by Queen Elizabeth II at Calder Hall in the UK. It is still in operation.

1957 The famous Zeta thermonuclear-fusion device was inaugurated at Hartwell, England.

1968 First turbo-alternators of 660MW each installed in the UK.

1972 First turbo-alternator of 1250MW installed in the USA.

Prototype Fast Reactor (PFR), rated at 250MW, commissioned at Dounreay, Scotland.

1982 The TFTR large-scale thermonuclear-fusion experiment started operating in the USA.

1983 The JET large-scale thermonuclear-fusion experiment started operating at Culham in the UK.

Glossary

AC Alternating current, which changes direction regularly.
Alternator See *generator*.
Ampere A unit of electric current, or rate of flow of *electrons*. Named after the scientist who discovered it. Abbreviated as Amp or A.
Battery An arrangement of *cells* to increase voltage or current capacity.
Cell A device which produces electricity by chemical action.
Charge Atoms can become charged by gaining or losing electrons. An electric charge may be of two kinds: positive and negative.
Circuit An arrangement of *conductors* by means of which current is able to flow. A simple example is the connection between the electricity supply and a bedside lamp via a *switch*.
Circuit breaker Automatic safety switch which breaks a *circuit*. It takes the place of a *fuse*.
Conductor Any material which allows electricity to flow. There are good conductors such as metals, and poor conductors such as water.
Cycle Time taken for one complete alternation of AC current (see *frequency*).
DC Direct current, which flows in the same direction all the time. A *cell* or *battery* produces DC.
Distribution The final section of a grid system where current is taken from a grid supply point and distributed to consumers at suitable voltage levels.
Electron The smallest particle of matter.
Fission Splitting of heavy atoms of uranium or plutonium to release energy.
Frequency The number of *cycles* of alternating current per second. The standard frequency in the USA is 60 and in the UK 50.
Fuse A safety device – either wire or a cartridge – which melts when too much current is used and breaks a *circuit*.
Fusion Joining together of light atoms such as deuterium and tritium to release energy.
Generator A machine which produces electricity. It is sometimes known as an Alternator (which only generates alternating current).
Grid system An arrangement of power lines connecting power stations and consumers over a large area.
Inductance When the current through a coil changes (increases or decreases), the consequent change in the magnetic field will induce a current into the wires of the *same* coil. This in turn produces a magnetic field, and so on.
Induction (1) *electrostatic* – process of charging an object electrically by bringing it near a charged object. (2) *electromagnetic* – the action by which a changing magnetic field induces current.
Insulator A material which does not conduct electricity, such as rubber, plastic and enamel.
Ohm The unit of electrical resistance. Symbol Ω (Greek letter for omega).
Ohm's Law The relationship between electrical voltage, current and resistance, expressed as Volts = Amps × Ohms.
Parallel Term used to indicate how devices are connected in circuits.
Power The term used for the product of voltage and current. It is measured in *watts*.
Power station The central station where electricity is generated for use over a wide area.
Pumped storage A special arrangement whereby electricity is used during a period of low demand to pump water from a low level to a higher level, where it is stored until needed. The water then flows down and drives turbines to create electricity.
Resistance All materials offer a certain amount of opposition to the motion of electrons. This action, which reduces the current, is called resistance and is measured in *ohms*.
Series Term used to indicate how devices are connected in circuits.
Series/Parallel An arrangement of cells which uses the *series* method to increase voltage and the *parallel* method to increase current capacity.
Switch A device which controls the flow of electric current in a circuit.
Transformer Device used in AC to change one voltage to another.
Transmission The transfer of current from a power station to a grid supply point which could be hundreds of kilometres away.
Turbine A motor driven by steam or gas used in power stations.
Turbo-Alternator A combined *turbine* and *alternator*.
Utility See *power station*.
Volt A unit of electrical voltage. The abbreviation is v.
Watt A unit of electrical power. The abbreviation is w.
Wind-turbine A wind-driven generator.

The energy that we use

The use of oil as a fuel is diminishing in all countries except those whose economy is based on oil production. In the USA some utilities are being converted from oil to coal. Coal and uranium are therefore the most important fuels except in those countries with enormous water power.

The proportionate consumption of fuels in the USA and the UK is approximately as follows:

	oil or coal %	nuclear %	hydro %	others %
USA 1981	76.6	12	11.4	negligible
UK 1982–83	85.9	14	0.1	negligible

In 1970 the forecast for 1980 for the USA was 19 per cent nuclear. It turned out to be considerably less – which shows that figures must be treated with caution. The nuclear proportion in the USA and the UK is small compared with countries such as France, where in 1982 it was about 40 per cent; the forecast is about 80 per cent for the year 2000, but much may happen to prove this forecast wrong.

Total electricity production

USA 1981   2,294,786 gigawatt-hours
UK 1982–83   234,519 gigawatt-hours
(1 gigawatt = 1000 megawatts = 1,000,000 kilowatts).

The six biggest consumers of electricity are: 1. United States of America; 2. Japan; 3. Canada; 4. West Germany; 5. France; 6. United Kingdom. (The USSR has been excluded because of the difficulty of obtaining accurate figures.) Taking the population into account, however, the three biggest consumers (per person) are: 1. Norway; 2. Canada; 3. Sweden. This suggests that electricity is the most suitable form of energy for supplying heat.

Who makes electricity?

Architects
Biologists
Boiler operators
Cable joiners
Chemists
Clerical staff
Computer experts
Control engineers
Crane drivers
Electrical engineers
Instrument mechanics
Linesmen
Maintenance engineers
Managerial staff
Mechanical and electrical fitters
Metallurgists
Physicists
Planning engineers
Power station operators
Quantity surveyors
Shift engineers
Substation operators
Wayleave operators

Index

Acid rain 8
Alternating current 12, 13, 29, 30
Alternator 13, 29, 30
Ampere 9, 30
Ampère, André-Marie 12, 29
Atom 4, 5, 18, 19, 28, 30

Battery 5, 9, 25, 26, 27, 29, 30
Boiler 6, 8, 9, 10, 11, 22

Cell 2, 17, 26, 27, 29, 30
Charge 2, 3, 4, 5, 30
Circuit 20, 24, 25, 29, 30
　breaker 24, 30
Coal 2, 6, 8, 9, 10, 11, 13, 18, 19, 21, 28, 31
Conductor 2, 5, 12, 20, 24, 30
Current 2, 5, 6, 8, 9, 12, 13, 17, 20, 21, 24, 25, 26, 27, 29, 30

Diesel 6, 9
Dinorwig 15, 23
Direct current 13, 29, 30
Distribution 21, 23, 30

Electromagnet 12, 29, 30
Electron 3, 4, 5, 30

Faraday, Michael 12, 20, 29
Fission 6, 18, 19, 28, 29, 30
Frequency 13, 30
Fuse 24, 30
Fusion 19, 29, 30

Gas 6, 28
Generator 6, 7, 9, 10, 11, 12, 13, 14, 15, 16, 18, 22, 25, 29, 30
Geothermal energy 6, 17
Grid system 7, 20, 21, 23, 30

Hydroelectricity 6, 14, 21, 28

Induction 4, 20, 29, 30
Insulator 4, 5, 20, 25, 30
Itaipu 14

Lightning 2, 3, 23, 29
Lines of force 12, 20

Magnetism 12, 13, 29, 30

Neutron 4, 5, 18, 29
Nuclear fission 6, 18, 19, 29, 30
　fusion 19, 29, 30
　power 6, 8, 9, 10, 18, 19, 21, 31
　reactor 6, 11, 18, 19, 28, 29
　waste 19, 28

Oerstad, Hans Christian 12, 29
Ohm 9, 29, 30
Oil 2, 6, 13, 21, 28, 31

Power station 6, 7, 8, 9, 10, 11, 14, 15, 17, 18, 19, 20, 21, 22, 23, 29, 30
Pumped storage 15, 23, 30

Pylon 6, 8, 20

Rance 14, 15

Solar energy 6, 17, 28
Static electricity 3, 4, 5, 8, 29
Steam energy 10, 11, 18
Substation 6, 8, 20, 24

Tennessee Valley Authority 14
Thomson, Sir Joseph 4, 29
Thunderstorm 2, 23
Tolerance 23, 30
Transformer 6, 7, 12, 13, 20, 21, 24, 29, 30
Transmission 6, 7, 13, 20, 21, 23, 29, 30
Turbine 6, 7, 8, 9, 10, 11, 14, 15, 29, 30
Turbo-alternator 14, 29, 30

Uranium 18, 19, 29, 30, 31

Van de Graaff machine 3
Volt 3, 9, 20, 30
Voltage 6, 7, 8, 13, 17, 20, 21, 23, 24, 26, 27, 30

Water energy 2, 6, 8, 9, 14, 15, 28, 31
Watt 9, 30
Wind energy 6, 14, 16, 28
Windmill 10, 16
Wind-turbine 16, 30
Wiring, domestic 24, 25

Acknowledgements

Threshold Books and the publishers gratefully acknowledge the help given by the Central Electricity Generating Board in the production of this book.

Illustration credits
Photographs: Central Electricity Generating Board, London pages 5, 8, 10, 11 (middle), 13, 15 (bottom), 16, 20, 22 (bottom), 23; Bruce Coleman/R K Pilsbury 2 (top), P A Hinchliffe 2 (bottom); Geoffrey Drury 26; Electricity Council 27; GEC Turbine Generators Ltd 11; John Hillelson Agency/C Simonpietri 7, Pierre Vauthey 17 (top); JET Joint Undertaking 19 (top); David Lockwood 21; National Nuclear Corporation Ltd 9; Science Research Council 3; Sodel Photothèque EDF/M Brigaud 15 (top); Solarfilma SF 17 (bottom); UK Atomic Energy Authority 19 (bottom); US Department of Interior 14; ZEFA Picture Library 12.

Diagrams and drawings: Ray Burrows pages 4 (bottom), 5, 11, 12, 13, 14, 15, 16, 18, 20/21, 22, 24, 26, 27; Gillian Newing 6/7, 28; Carole Vincer 3, 4 (top), 25, 29.

How Electricity Is Made
© Threshold Books Limited, 1985
First published in the United States of America
by Facts On File, Inc.,
460 Park Avenue South, New York, NY 10016.
First published in Great Britain by Faber and Faber Limited.

All rights reserved. No part of this book may be reproduced or utilized in any form or by any means, electronic or mechanical, including photocopying, recording or by any information storage and retrieval systems, without permission in writing from the Publisher.

General Editor: Barbara Cooper
The How It Is Made series was conceived, designed and produced by Threshold Books Limited,
661 Fulham Road, London SW6.

Library of Congress Cataloging in Publication Data
Boltz, C. W.
　How electricity is made.
　Summary: Describes the various sources of electricity; how it is harnessed, controlled, and generated; and how it is distributed to provide us with power and light.
　1. Electric power-plants——Juvenile literature.
2. Electricity——Juvenile literature. [1. Electric power-plants. 2. Electricity] I. Title.
TK1191.B64　1985　621.3　84-21050
ISBN 0-8160-0039-5

Typeset by Phoenix Photosetting, Chatham, Kent, England
Printed and bound in Belgium by Henri Proost & Cie PVBA